I0423107

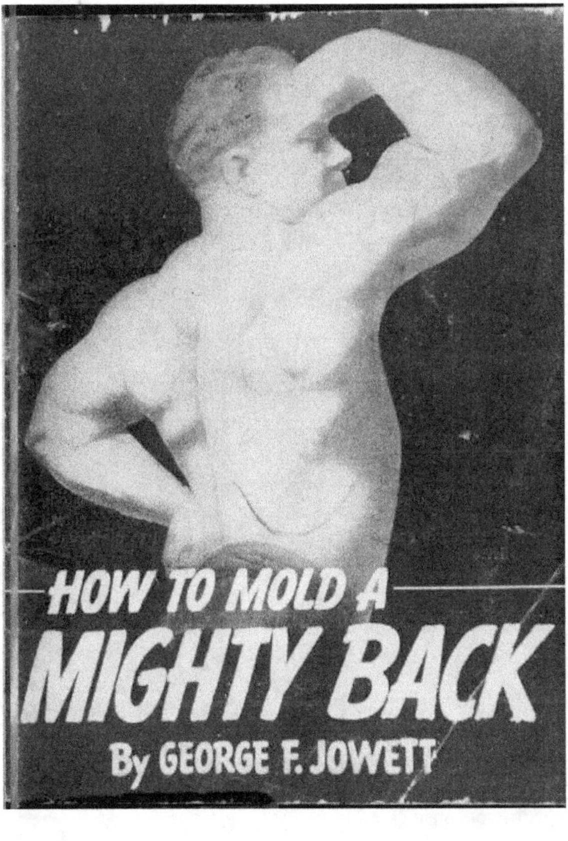
HOW TO MOLD A
MIGHTY BACK
By GEORGE F. JOWETT

HOW TO MOLD A MIGHTY BACK

(ORIGINAL VERSION, RESTORED)

By

GEORGE F. JOWETT

Original Publisher: The Jowett Institute of Physical Culture, 230 FIFTH AVENUE, NEW YORK, N. Y. 1938.

PUBLISHED BY O'Faolain Patriot LLC, Copyright 2011

info@PhysicalCultureBooks.com

ISBN-13: 978-1466476714

ISBN-10: 1466476710

Published in the United States of America

To Order More Copies Visit: Physical Culture Books.com

JOWETT BREAKS RECORD
of the Strongest Man IN THE WORLD!

…Yet he was once a physical weakling.

On the cover of this book, you see a picture of George F. Jowett. Look at his mighty back and his terrific arms—the strongest arms on record! It was claimed that Arthur Saxon's one arm record would never be broken. It stood through the years, till Jowett officially broke it and established a new mark not since equalled! Jowett holds more records for strength than any other man living or than any other man did at any one time. Because he holds more world championship records than anybody else, he has been called "Champion of Champions", and because he has built more world champions he is called by experts, "Builder of Champions".

ENROLL NOW for the famous Jowett Course, and build yourself in the way he made himself into an unsurpassed giant of strength, out of a physical wreck condemned to die before fifteen!

MOLDING A MIGHTY BACK
By GEORGE F. JOWETT

WHENEVER the memory of my father comes before me I always visualize his physical proportions which were accentuated by the great breadth of his shoulders. Few men have had such a massive spread across the broad of the back and it was this more than anything else which drew my admiration to him when I was only a young boy. There is something about a broad back that immediately arrests the attention of every eye. It seems to proclaim so expressively the radiant health and great strength the owner must enjoy. A man may have good arms, chest and legs, but his clothes rob others of the opportunity to appreciate the development of those parts of his body. Not so the shoulders, they loom before the sight of all, displaying their splendid back formation, or undeveloped ugliness, as the case may be. Naturally, we recognize the importance of a brawny back, so much so that we frequently employ it in our daily vocabulary in such expressions as "Put your back to the wheel," "Shoulder your way through to success," and "Man, but he has backbone," and these expressions in each case are singularly true.

The development of a mighty back, first for myself, and later for others has been one of my hobbies. Before we begin on yours I want to be sure that you are fully acquainted with the muscles of the back so that you will understand all that I say to you.

In Fig. 1 we see the two latissimus dorsi and two trapezius muscles. The first named mean "big, broad back muscles," forming the bulk of the back. Originating from the spine they taper into a powerful ligament which becomes strongly fastened on the upper arm.

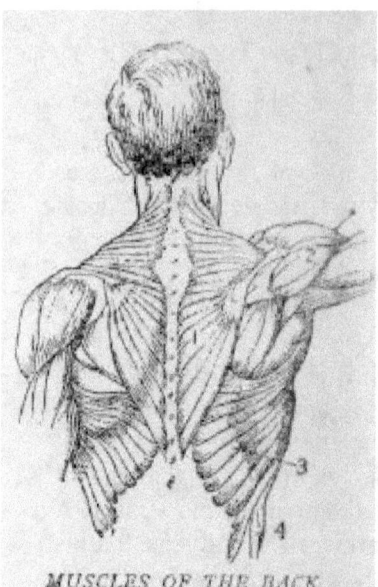

MUSCLES OF THE BACK
1. The trapezius. 2. Back deltoids.
3. Latissimus dorsi. 4. External oblique.
Note the varied directions in which the muscle fibers run. The diversity of direction is what gives the muscles their single, dual or triple volition.

To all weight lifters these muscles are highly interesting. Eugene Sandow was the first to properly understand them. He realized what a splendid shelf each formed for the upper arm to rest upon as the body bends sideways in the opposite direction. It actually supports or props the arms into straight arm rigidity. From understanding the latissimus dorsi the famous lift known as the "Bent Press" was born.

Some call the "Bent Press" a trick lift or feat of balance, but that is foolish. If the latissimus is not developed strongly enough all the balance technique in the world would not make a successful bent press lifter. Skill in balance is an asset, but power is the fundamental, though we might safely say the big back muscle lifts or pries up the weight more than does the arm. Arthur Saxon was the greatest demonstrator of latissimus efficiency, raising to arm's length, officially, a total poundage of 371 lbs., and unofficially 4091 lbs.

The trapezius is a peculiarly shaped muscle. It arises from the spine in a long taper then triangles out to fasten on the shoulder blade. From this connection it runs back up the neck and finally becomes attached onto the base of the skull. This muscle is capable of triple action. By contraction it helps the latissimus to aid the serratus magnus muscles to deepen the
chest; it also pulls the shoulder blades back. When performing this duty it displays contraction a little differently by bunching up the muscle at the base of the neck into a saucer like depression.

This shows you that the most powerful volition of these two back muscles is in contraction and that this action shortens the back muscles. This shortness is what deepens the chest, squares and *#* spreads the shoulder. For these reasons the back muscles are important in their relation to the chest.

The trapezius muscles are responsible for the erect carriage, and the latissimus dorsi for widening the shoulders, two very important points which should be kept foremost in your mind.

The deltoids are those mounds of muscle which cap the shoulders. The word is derived from the Greek word "delta" because the muscles have the formation indicated by that word which is much like a triangle. This muscle is formed by three overlapping leaves of muscle which are spread out over the shoulder from where they originate. They taper off in wedge shape into the arm muscles. These muscles assist the arm in all lifting movements and control the rigidity of the arms when they are held out sidewise in a line level with the shoulders.

The external oblique muscles mean the outside oblique shaped muscles. They support the outer edge of the latissimus dorsi muscles at the side of the body, and roll over the side forming the outside edges of the basin which contains the abdominals to which they are joined by a grisly flat sheet of tissue. They control all side body bending movements and protect the groins against hernia.

Altogether you have now gotten a fairly comprehensive picture of the muscles in which we are most interested.

And now, while the thought occurs to me, if you ever get the opportunity to study the likeness of the back development of the great George Hackenschmidt, don't miss the chance to give it a deep, searching scrutiny. There is so much to admire in his Herculean, picturesque back—more so than in most of the back development shown by the great men of physical power, except of course the back of the equally great Arthur Saxon. THERE is revealed to you all the magnificence of muscular grandeur, which shows the possibilities of the right type of exercise and what can be brought out in muscular back development. From his hips the small of the back rears square, then flares up and out in wedge shape into a massive fan-like flange merging under the arm pits. Here the shoulders provide a gigantic ledge for his powerful neck and shoulders that sit squarely upon this solid foundation. The remarkable sculpture of his every back muscle is amazing. Each muscle is so clearly defined it is a separate unit, but under balanced control. Your enthusiasm will mount higher as you notice the roll of the trapezius muscles as they mould into their great cuplike formation at the base of the back between the deltoids.

Perhaps the most beautiful of all are the two long rope-like columns which form the erector spinus (straighteners of the spine) which flank each side of the spine from the neck down to the pelvis. In the small of the back these muscles display their magnificent development most powerfully, giving the appearance of corrugated reefs into which the back bone is firmly embedded. Those who were fortunate enough to see him in his prime will recall how, pythonlike, those two twin columns would twist and twine with the bend and sway of his every back movement, bolstering the spine at every turn with twin walls of flexible, grfnite-powered muscle.

Every body builder has an ideal. Quite a number of friends of my boyhood days claimed the classic Eugene Sandow, but two men aletife ruled my physical dreams—George Hackenschmidt and Arthur Saxon. If any person came nearer than I to idolizing those human beings they went some. I almost worshipped those two men and I owe all I have gotten in strength and development fo jfchem. They started me at the point where I began to see the fjrfet gleam of success. They guided me when I came to the point of perplexity. They were my ideals and my inspiration.

Hackenschmidt had a figure every bit as classic as Sandow, but more Herculean. Literally, he was twice as strong as Sandow, and far more versatile. Sandow was a poor wrestler.

Saxon was a different type. He lacked the physical beauty Hackenschmidt had, but for strength he was superior to the Russian Lion.

No doubt you have an ideal and see in him what I saw in these two giants of my inspiration. You, no doubt, hope to acquire a build equal to that of your ideal, but without that wedge shaped back you will never be physically perfect.

No matter what your body weight may be you can possess a magnificent back. A man does not have to be a heavyweight in order to possess the shoulders of a grenadier, though I know every aspirant to physical perfection longs to become a heavyweight. It has been a never ending question as to why two men reared under the same conditions reach manhood in different physical states. In this there is an underlying problem which neither I nor anyone else can fully answer. In my boyhood I was very small. Many of my comrades were much larger than I, yet one thing puzzled them greatly—despite the physical inferiority of my stature, I was stronger than any of them and much quicker. As we grew up to manhood, I became the largest of them all even though some were taller. Some of those who were big for then- age, as boys, became undersized as men. Several of them were devoted to physical culture, but none of them ever made the gains I did. Of course, I know now, much of their trouble was lack of right instruction. The fact that I was later so strong as a youth brought me in contact with many notables and I was fortunate in receiving much valuable instruction from them.

There are natural reasons to explain why one grows large and others remain small, reasons that I have studied all my life, and I have been singularly successful in arriving at many correct conclusions. I know that the back above all other parts of the body, including the chest, has the greatest direct influence upon increasing the energy, power and size of a person and for the one chief reason that within the bones of the spine is encased the electricity of life—the spinal cord. From neck to hips there is, throughout life, a continual increasing vital life going on here to intensify the organs and muscles. Due to the nearness of the back muscles to this wonderful source of invigoration they have the better chance to obtain more dynamic nourishment for building larger muscles. One young fellow asked me, "Why don't these muscles grow without exercise when they get all that stimulation V* He overlooked, like so many others do, the fact that muscles grow only from use. That is why a blacksmith is stronger than a person who merely does clerical work. Pushing a pencil does not require the physical effort the swinging of a hammer does, but even the most strenuous occupation fails to give a person the muscular development you and I admire. The muscles only grow to a certain size, enough to supply the need, then they cease to grow. After that the muscles begin to acquire endurance for the daily task—which minimizes the demand upon the muscular strength.

If people had to swing hammers, or lift around pianos, in order to become strong, I don't think any of us would ever be very strong. We all have a natural antipathy for strenuous work, but there is a world of difference between work and exercise. In the first place exercise is cultivation, and a little of the right kind goes a long way—leaving you feeling fresh and full of energy. The answer to this is that daily, commonplace movements rarely furnish one with complete muscle contraction and extension, while proper exercise is a complete series of both. I know a very powerful blacksmith who told me that after a hard day's work at the anvil, he was more careful to do exercise than on easier days as it refreshed him so.

I asked him what form of exercise he preferred and he told me "back movements," the kind that call for bending forward and slightly backward, loosening up the muscles that surround the small of the back. Incidentally, it is there that is located both the weakest and the strongest part of your spine. Weakest in the fact that the spine has no other musoa-lar structure to fall back on between the hips and rfbs except what naturally exists. Consequently, this important and mud* abused bone requires protection and muscular support a Ihtls more than 4oes the rest. Every bodily move rotated from above the hips calls for a ceaseless movement of the spinus erector mus«l«s, and the muscles of the side we call the external oblique. It is the strongest part of the spine for the amount of muscles that subpart rt and is capable of absorbing much more physical abuse throughout life than any other muscular group in the body. Even in an undeveloped state the spinus erector muscles are enormously powerful. If you clasp your hands behind the neck, then throw the elbows back, as you tense the muscles you will feel a decided tightening of the muscles of the upper back, but if you will throw the hips forward and bend slightly to the side, you will feel a more powerful contraction in the small of the back. You will feel the spinus erector muscles straightening the spine. If your back lacks development in this part there will be nothing appreciable to see, but, if developed, the spinus erectors will rise prominently showing a detp cleft running through the middle as shown in Fig. 2.

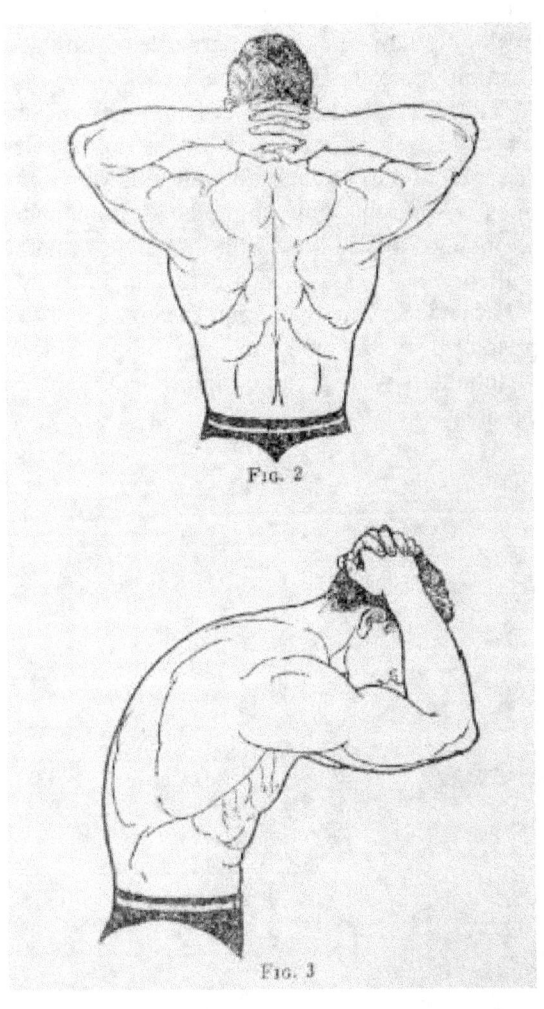

FIG. 2

FIG. 3

As long as you have a well built pair of spinus erectors you need never have any fear of developing lumbago, kidney trouble, or any other affection in the small of the back.

To get a different view of the back muscles you have only to draw in the abdomen, lean forward and pull down with your elbows with the hands clasped this time on the head. Immediately the mounds of muscle will flatten out and the small of the back will cease to arch; instead it will round out so one long sweeping curve is apparent from the hips to the base of the skull and the shoulders will spread out fanwise as shown in Fig. 3.

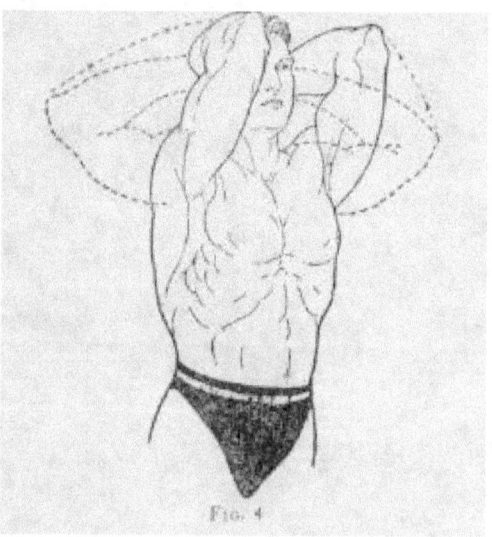

FIG. 4

This little stunt will show up your defects. If you are short on back structure the half moon curve will not be visible. Instead, there will be a flat spot back of the waist region showing the points of the spine sticking up pressing against the skin. At the base of the shoulders where tike neck arises there will show up a bony lump and the neck will have a hollow, scooped-out look. There will be no "shoulder spread."

To get the "feel" of the big back-spreading muscles all you have to do is to raise the elbows out sidewise in a line level with the shoulders, then tense the muscles and bring t^e elbows together in a dosing forcep movement, pulling down forcibly with the nands on the head. See Fig. 4. You will feel the muscles of the upper back pull as they cause the shoulders to spread.

You will he surprised how much this pulling down wing-flapping movement, used as an exercise, will help to broaden the back and increase the size of the chest Gains of 4 to 6 inches are not tra» common. This is purely muscular gain and gotten from the growth of the latissimus dorsi, the big, broad, back muscles, that form the bulk of the back and which are attached mainly to the greatest length erf the spine. They insert into the upper arm on the under side of the humerus bone close to the shoulder arm muscles, by a ribbon like attachment which is powerfully seated on the humerus bone.

All movements that call for raising the arms overhead, pulling and shoving and spiraling the body on the waist, arc governed by the latissimus dorsi muscles. They do not need to be developed much in order for you to get the idea of their formation. They jut from the side, triangle

17

fashion, providing a formidable ledge for the upper arm to rest upon in certain feats of physical movement. This muscle is seen on muscle builders in varying sice according to their development. Some are capable of displaying enormous "shoulder spread" and muscular size but frequently they are not as strong as you would expect in proportion to their muscular size. Few strong men who cultivate the lifting of heavy weights, as a means to building up the latissimus dorsi muscle^ acquire the power that it is possible to get from these muscles. This is proved by the general inability of these men to raise a weight overhead of any worth while amount m two separate dumbbells or to raise an appreciable amount of weight in the one arm lift known as the "Bent Press" which I have already explained in this talk.

Gymnasts who specialize on the Roman rings, hand balancers and rowers acquire the finest looking latissimus dorsi muscles. Weight lifters, who are good at separate dumbbell lifting and the '*Bent Press" have splendidly formed back muscles but as there are so few of them we will not consider weight lifters for the time being. The first named athletes, that is, those who specialize on the Roman rings, naturally acquire powerful latissimus muscles because every stunt they practice calls for a great pull or thrust movement from them.

Undoubtedly these latissimus dorsi muscles are the most influential in providing you with a big broad back. They give you a "flat back," not a back which, viewed from the^ side, shows the deep bend inwards in the location of the waist region. I have often seen this inverted arch erroneously referred to, by some uninformed writers of physical culture, as the curve of beauty and sign of strength. Right here lies the unfortunate result of training too soon with heavy weights. The constant raising overhead of a heavy bar bell, or even of one within your limit, done frequently and with too many repetitions, causes the back to sink in. This is

Fig. 5

not a sign of development It is a proof of weakness and becomes what the medical profession and chiropractors call an inverted spinal curvature—an abnormal back condition or deformity. The muscles not being strong enough in this back region causes the spine to be forced

19

forward. The result is an inverted curvature which brings on backaches and a weakness which causes trembling spells after a lift. Unfortunates, so affected, are never able to sit on a chair correctly or even walk right. When walking, the abdomen is carried forward too far and the back soon tires. When th^f sit down the back slumps or folds up like a weak hinge. Many fine, splendid phvsical specimens have been ruined with this foolhardy practice and it is all the fault of the people who advise this form of training without providing the pupil with the right preparatory and careful thought, advice and instruction.

As a matter of fact the old exercise we call "back bending" is » not so good, not unless your back is, in the first place, safely and strongly built. At any time, the right way to practice this movement is not with the hands cupping the waist but with the palms of the hands placed flat upon the hips. See Fig. 5. This forms a support and will safeguard the back against strain if you go too far back. The latter method provides a Dolstering support whereas the oldtime method caused a back strain in order to recover an upright position.

Strangely enough "back bending" as an exercise has a peculiar effect on the nerves. A few far back movements will make the muscles in the small of the bade quiver and a strange top-heavy, wobbly sensation is left that makes a fellow feel he wants to relax, by sitting or lying down. On the other hand, if you bend back slightly, say about six inches, there is sufficient

20

back muscular stimulation gotten for the small of the back which is just as astonishingly beneficial as bending too far back is detrimental. As a matter of fact actual "side bending" (without any forward inclination) will develop the spinus erector muscles more quickly and safely.

Siebert, the great German authority, warns against the "back bending" practice, and Arthur Saxon was the one who advised me as a boy to "cut it out."

Well, let us go back to the latissimus dorsi muscles again.

I suppose you have often "chinned the bar" or, at least, tried. Did you ever allow yourself to hang suspended by the hands from the bar for a few seconds? If you did you will recall the pulling sensation experienced under the arms and on the back. I knew a body builder who claimed hanging by the hands from a bar daily, gave him an increased shoulder spread and chest increase of nine inches. This does not mean he actually made his rib box grow nine inches, but the latissimus dorsi muscles grew, thus greatly increasing his shoulder spread and broadening his back so that he totaled an accumulation of nine inches by the tape measure over his original chest girth.

Max Sick, the marvelous Bavarian athletQ, practiced most for back development what hand balancers call "press outs." I happen to know this exercise was one of his pet exercises. So much did he believe in it that it finally became his only exercise (after he had gotten his development). It certainly is a wonderful exercise if you are able to do it. The idea is to stand on your hands, then by bending the elbows, lower the body until the chin almost touches the floor and from that stage press yourself back to arms' length in the handstand position, doing it several times. Perhaps you are not able to make a hand stand and control it to ao tncse "dips" and "press outs." In that case stand on your head a few inches from the wall and throw your feet up against the wall. A little practice will enable you to press your body weight to arms' length until in time you will be able to do it as Max Sick does. T used to practice this stunt on the ground bars or between the parallel bars. The latter is easier and in fact gives you more shoulder and back action. Roman ring practice is not so easy. Many people are never able to do what is known as the "full mount". The heavier your body weight, the less chance you have to ever do it. Roman ring artists are generally very light and are mostly short men, with very little leg development which makes their task easier.

One of Arthur Saxon's favorite back exercises was to raise to single arm's length a dumbbell, well within his power, and while holding the dumbbell at arm's length, bend over sideways. You will find the bending over is the easiest part, the straightening of the body is what gets you. So do not be too ambitious at the start. Stand with the feet about eighteen inches apart, though there is no set rule as to how wide apart the feet should be, as long as you feel it is sufficient for your comfort and balance. Keep the arm as straight as possible and bend from the waist sideways. See Fig.6. If you bend a little forwards, it will help, as your pelvis or hip bone is apt to limit your bend if you try to make a perfect side bend.

Fig. 6

Of course, you realize you must bend to the opposite side of the erect arm, and do not forget to perform the practice with the left arm as well as the right.

Now don't ask me how much weight you should commence with. That is something for you to decide upon. At first you may find it a little difficult to control your balance, which for the time will make the exercise a little harder, so the lighter the dumbbell is the better.

Saxon preferred the dumbbell over the bar bell for this exercise for two reasons. First, a dumbbell is harder to control and so compels more of a stationary position, throwing proper resistance upon the right muscles. Second, because of being harder to control, you are not able to handle as much weight, and so remove the possibility of trying to handle more than is beneficial for the exercise.

Here is an exercise that is another favorite of Saxon's and is by far one of the finest I know of to give any person a powerful back.

You take up your position with the arms held overhead, with a disk dumbbell held in each hand. The feet should be spaced apart, and the whole poise must be one of balance. From this position, you start.

Breathe in and bend over SIDEWISE as far as you can. Carefully examine Fig.7, before you begin. When you have gone the limit of the side bend come back to your first position and bend over to the opposite side. The big feature in this exercise lies in HOW far you can bend. The arms must never lose the original angle and at all times must be kept in a direct line with the body bend. Don't bend the least bit forward.

Fig. 7

Back development with Saxon was a thing of vast importance. He rightly claimed that too many would-be strong men spoil their possibilities by placing too much reliance on the arms. Your back is stronger than your arms and if you understand the science of handling heavy objects, you soon find this out Saxon raised more weight overhead, single handed, than any other man has ever done using both arms. As I have already related he officially rilsed overhead with one hand, in the lift known in the sport of freight lifting as the "Bent Press," 371 lbs., and unofficially 409! lbs, It really seems a performance too stupendous to believe, but it is a fact. Nevertheless, it was only made possible by the terrific back strength this mighty man controlled. I say controlled because there is all the difference in the world between muscular control and muscular possession. Many people possess great back strength but cannot control it. Saxon could control his muscular mechanism with the same exact precision and confidence with which an expert mechanic does the levers of his machine.

The "Bent Press" is purely a feat of back strength. The arm is merely a prop with the seat of balance resting within the hand. Success in this feat rests with the person knowing how to place the arm on the side and KEEP IT THE&E throughout the lift and knowing HOW and WHEN to bend the knee of the non-lifting leg. The strength required is supplied by the latissimus dorsi muscle to provide a granite-like ledge and a thrusting power with which to straighten the arm. The spinus erectors plays an important part as do the external obliques in supporting the giant latissimus dorsi as the body bends, twists at the waist and finally becomes erect.

The actual performance of this great lift is to stand with the feet a well balanced distance apart and with one or both hands lift to the right. shoulder a loaded barbell or dumbbell. The body is curved to the left a little and the right elbow is placed firmly on the side of the body within the fold of the latissimus dorsi. (Don't try to look for the fold. If you have it the upper arm rests there naturally.)

The right leg must be perfectly straight and the lifting forearm in a straight line with this leg. The hand is twisted so the globes or discs of the weight are in a line straight with the shoulders. From this position you begin to SPIRAL the body from the waist to the left side allowing the left arm to slide across the left thigh in order to steady your balance. The position is clearly shown in Fig. 8. It is important that the legs should not bend until you feel you have bent over as far possible, then you bend the left knee and "GET UNDER" the weight. Throughout the performance you push steadily with the right hand against the weight. When the arm becomes straight, which the bending of the left knee brings about, you rise erect. While there are other scientific methods of performing this feat, this is the general method, the other KINKS you naturally work out from practice. The thicker the bar of the weight is, the better for this feat. It provides more pushing space. This is important.

Fig. 3

As I have already said, this feat is performed mainly by the latissimus dorsi and the very fact that this muscle is capable of compelling the arm to elevate a poundage beyond the possibility of the muscles both arms together are capable of, gives you an idea of the possibilities of its gigantic physical resources, when properly developed. As they build up in structure a great flatness across the back becomes evident, and this creation of immense breadth also widens out the shoulders.

A splendid exercise to develop the muscles that govern the "Bent Press" is illustrated in Fig. 9 which shows the position of the exercise when half way through. The next half, coming erect, calls strongly into play the muscles of the side. The arm behind - the back is the simple part that makes this exercise superior to any arm position. Be sure to keep the lifting leg straight, only bend the other. You will find this exercise a dandy in more respects than one.

Here is a peach for giving the entire back a workout in contraction and extension. The first time you practice it you will feel a sensation upon the breast bone and in the shoulders akin to spreading apart It will make the cartilaginous structure of the costal region wake up a little more.

FIG. 9

Take up your position as shown in Fig. 10 (a), with the bells placed between the feet Be sure the feet are spread a comfortable distance apart With a mighty heave tear them off the floor, breathe in, and hurl them to arm's length overhead and continue by allowing the arms to fall down and out as in a crucifix position.

In this exercise you really describe a complete circle by raising the weights off the floor, and throwing the arms back to give the chest and shoulders all the possible spread.

Fig. 10 (a)   Fig. 10 (b)

Do not bend the arms while this exercise is in process and practice it as vigorously as possible. Study the illustrations carefully, and note how the head is thrown back as the downward movement commences so that the chest spreading effort is aided the most. Do not move the feet. As the bells descend to the floor between the feet prior to making the next movement, breathe out; breathe in as you repeat the circling movement.

Fig. 10 (c)

If it were possible for you to examine the tissue of the latissimus dorsi you would be puzzled to note how the fibrous threads of muscle that form its bulk vary in direction. Now the muscles on the front of the upper arms, called the biceps, are constructed of a one way fibre, they are straight because this muscle controls the arm only in one movement—the bending of the forearms on the upper arm; but by this time you have realized the many movements the big broad back muscle is capable of. It is a contortionist, which is the reason for its varied fibrous construction.

People who are round shouldered and narrow chested have very poor latissimus development and slump when walking or sitting; therefore, it is reasonable to assume development of these muscles will overcome those defects. Men with broad shoulders have a very erect carriage when walking or sitting.

Where a spinal curvature exists, the latissimus dorsi is drawn out of shape which is plainly shown by its twisted appearance.

By scientific exercise treatment I have been able to correct many bad curvature cases. Two of the best exercises to practice in this case are very simple. I have the pupil arrange a bar about two inches higher than he is able to reach overhead unless he rises on his toes. By grasping thi$ and merely hanging, the muscles on both sides will lift and pull—not all over—but particularly in the location of the curvature, because there the muscles are unnaturally stretched on one side and too much contracted on the other side. The action will help straighten the displaced vertebra? and help the muscles in that region to contract and relax and regain the natural condition.

The next exercise, which really is only .a progression on the first, is to Stretch until the toes are on the floor (see Fig. 11) and then try to force the heels down. Where the curvature is more pronounced on one side of the back, the object should be to grasp the bar with one hand and stretch until both heels are on the floor. If the pronounced curvature is to the left side hang with the right arm; use the left arm if the curve is to the right side.

I don't know whether you are familiar with the name of Bobby Pandour. Here was a man who possessed a marvelous development. One of his best poses is a natural position with the left hand grasping a bar suspended overhead. The effect it had, is to show the remarkable development of his wonderfully moulded back muscles. I only wish you could see that picture. Rarely does one see such superb back development. Of course there are others such as Kliment, Moygrossy, ana Arco. Then there are the mighty backed Saxon and Hackenschmidt who are considerably heavier men than the first three mentioned and of course they possess more imposing back development which makes them appear more impressive. The one thing I wish to emphasize on the back development of these men is the powerful squareness of their backs from the hsps to where the latissimus dorsi begin to flare out. Many display a back or shoulder spread pose which seems to flare out from the line of the Mps. Such backs are faulty and weak. Their "spread" does not mean they have powerful back development, or have large latissimus dorsi muscles. In fact it means the development in the lower back region is very deficient Look for the wide hips •with the square lower back ; in such you see * powerful muscular pillar of muscle firmly entrenched within the hips. Without this squareness, no matter how strong your upper back is, the back as a whole will be weak.

An important fact to remember is, within the area of your lower back, reposes all your virile power. That is why this part of the back is known to us as the sacrospinalus—meaning, the sacred part of the spine.   .

Do not believe that the only way to obtam great back development and power is to employ nothing else but exercise that creates only movements of great muscular contraction. You must stretdi the muscles also. The more elasticity and flexibility each musde has, the more power your back will have.

Perhaps you have noticed some of the heavy coil springs used on machinery and have observed the short compression some have for absorbing recoil and movements of depression. They are made for this purpose, of course, but what I am getting at is that of expansion they have none. Other springs have so much extension and compression they seem to

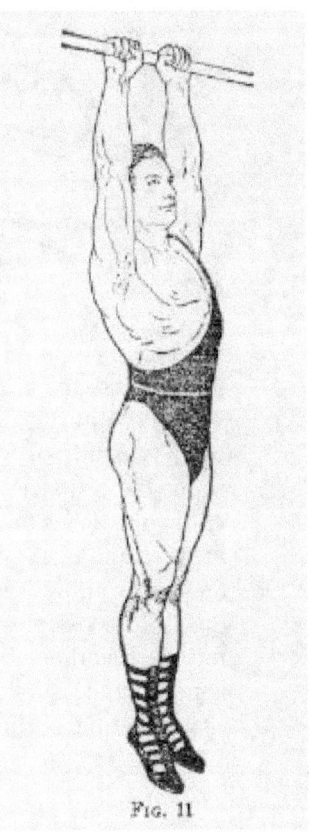

FIG. 11

35

be alive the way they are able to control and absorb the movements of extension and compression. That is the way your back muscles should be and for that matter every muscle in your body. There is more "punch" to them.

Fig. 12

Just try this movement. Sit on the floor and place across the soles of the feet a round stick, as shown in Fig. 12. Grasp this stick with each hand on the outside of each foot with the hands apart about the distance of the width of the shoulders. Allow the knees to be bent, then slowly straighten the legs and push on the stick with the feet. With the hands pull all you can and round the back as much as possible. Accentuate this feeling by further rounding the back and pulling with the hands on the stick. Then relax, bend the knees and repeat the performance several times.

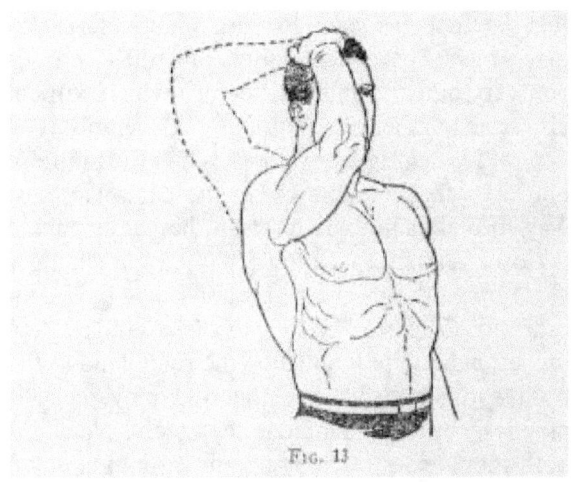

Fig. 13

Here is an exercise that combines two muscular controls, throws out and stretches the latissimus dorsi and actually contracts them. Stand erect and place the right hand on top of your head. Have the elbow pointing forward as in Fig. 13. Tense and force the elbow further forward, then move the elbow up and then out sideways until it is raised up in a line level with the hand. Repeat the movement several times, but do not forget to work the shoulders separately. After this you can practice the same movements with both hands clasped on top of the head.

The "Shrugging" exercises have always been popular for developing the trapezius and are among the best exercises to develop those lumps of muscle which form between the shoulders/ To me they are rather tiresome simply because so many repetitions are necessary before you really feel the trapezius begin to work. Of course, if you place plenty of concentration into the exercises you very quickly find a difference which will cut down the number of repetitions considerably.

A "Shrugging" movement is an effort to raise one or both shoulders up to the ears. The best way to practise this is to clasp the hands behind the back and endeavor to "lift" the right shoulder close up to the right ear as shown in Fig. 14. In order to get more "kick" into the exercise allow the left hand to pull down on the right hand as the right shoulder lifts. See that the movement is practised by each shoulder separately. Next, lift both shoulders up at the same time, pulling down with the hands in order to acquire irreater resistance.   .

Here is an exercise I always enjoy and one that I have practised all my life. In fact, it was one of the first group of exercises I ever practised. You may not care for it as much as some other, but I liked it because I could get tremendous resistance from it.

I first adopted the position of hands clasped behind the back, .⊬ then drew the shoulders forward and allowed the chin to sink upon the chest. I would then tense and forcibly bring the shoulder blades together as much as I could, throwing back the head as this was done. See Fig. 15. This I would repeat anywhere from twelve to eighteen times, giving complete contraction and extension of these powerful shoulder muscles.

The exercise brings to my mind a rather involved discussion upon exercise. You hear one say one method of exercise is better tikm another; one is for weights, another for no apparatus at all and another believes in using will power only. Then there is the leverage system, and so on until one gets so balled up he does not know whach the dickens is right.

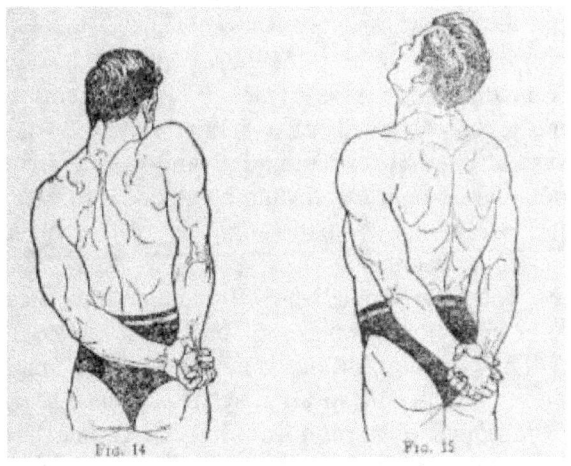

Fig. 14   Fig. 15

It is an admitted fact that I have a powerful pair of shoulders, and a broad back with a deep chest. Writers used to rave about my back and cheat development and yet the last mentioned exercise was always my favorite. You cannot call it a will power movement or one that strictly relies upon concentration. There is a great amount of physical resistance gotten from the arms and the ability of the deltoids to help force the shoulders back. When I > posed for my first muscular picture I chose this exercise which showed the trapezius cupped so impressively they formed a powerful ledge of muscle and though I was then only 16 years of age, my back development was said to be superior to that of many well- known muscle builders who had reached manhood.

Personally, I do not care too much for "will power" exercises or those which involve concentration unless they are combined with some other form of resistance. It always seems to me to be too much of a drain on the nervous system I am all for building up and would rather not do an exercise, no matter how good it is rated, if I see it causes an organic, nervous or other muscular reaction.

There are people who will mention Max Sick as a marvelous example of one who employed will power only. But this is not wholly true, and I think those people speak more from hearsay than from actual observation of this great athlete. I knew him and saw him train day in and day out. Most of his exercises he did practise without the use of apparatus, but in all of them he supplied a resistance on much the same manner as employed in the last explained exercise. He employed one group of muscles against the other. In other exercises he used light dumbbells. To some this may seem strange since his records and achievements as a strong man make the best living middleweight strong man look like a baby. His statement in his book is "I do not have to use heavy weight in order to accomplish great physical feats. Positions are the most important to learn but they cannot be learned struggling with heavy objects, which takes almost all of your strength. Neither can strong muscles be built struggling with heavy weights. Struggling with heavy weights keeps the mind from concentrating wholly upon the exercises and so robs the muscles of the benefit of the exercise." It is common sense. Arthur Saxon preached the same story and practised it He said •'When a man raises, once only, a heavy weight, all that he proves himself to possess is muscular control and great contractile power, but this does not guarantee sound internal organs nor does it prove that a man would come out well in an endurance Hest." He fur* ther states "Take care of the organs and they will take care of the muscles." I could repeat and produce proof upon proof of testimony believed and practised by other giants of strength and the substeuflfc of their teaming is all in the same vein, namely, exercise with heavy weights is not necessary for muscular development. Both Max Sick and Siucon stand without equal. Both were dsep students of the body and knew its physical mechanism from A t© Z. Both are living examples of what they practised and preached. Neither was ever influenced by commercial influences to sell or sign lying testimonials. I defy anyone to produce a writton testimonial to the contrary ever signed by either man. I quote what each wrote in the only book each ever wrote. I knew them intimately as I knew others. There was a time in my young life when I became so obsessed to lift,

I lost my better reasoning, and became so weak I could not lift half with two hands that I had been able to lift with one. I was seventeen years old at the time, and have to thank both Saxon and Hackenschmidt for correcting my method of training.

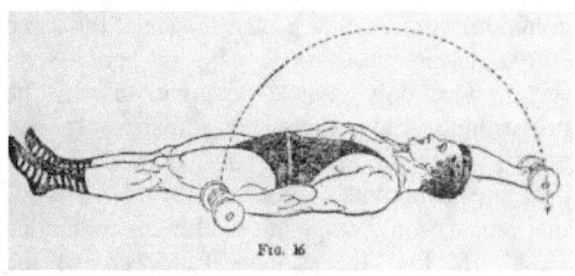

Fig. 16

The unfortunate thing in the body building training of many young fellows is that for some reason they easily believe things that are contrary to fact. Facts alone count and prove the substance of the effort. Follow the advice of the great masters who proved by their superlative health and rugged strength and physical perfection the truth of their beliefs and let nothing else deter you.

Baseball players often lose their pitching ability simply because they are afraid that exercise will tie up the muscles. TTiey have seen the results of faulty exercise training and are skeptical—but they need not be.

The trapezius and latissimus dorsi muscles control the secret of ball pitching and the right form of exercise will develop greater pitching force as well as preserve it, and safeguard the pitcher from acquiring an early "glass arm" which is a term that signifies tost pitching power. There is one exercise which, daily practised, will prove a blessing to the baseball pitcher. I have taught it some of our greatest stars who have proved its merit. In exercise instruction we call it the "one arm pull over."

The exerciser lies flat on his back with the pitching arm thrust to arm's length behind the head. See Fig. 16. In the hand he grasps a very light dumbbell, about two pounds, but not more than five pounds at the most. From this position raise the dumbbell over the head in a half circle movement with the arm held straight until it rests at arm's length down by the side, then return to the original position. Stretch the arm to full arm's length behind the head and keep it stretched to its full length throughout the exercise. As the performance is gone through the pull will be felt on both the trapezius and latissimus dorsi muscles. A similar pull the pitcher will feel if he pitches a ball slowly. Then, the stronger and more flexibly these muscles are developed, the more hurling force he acquires. In a similar way a boxer will develop a greater punch, though the boxer can add another valuable back exercise by punching straight forward from the shoulders with a light dumbbell held in each hand. This develops a more snappy latissimus forward drive.

The interesting part in back moulding, as I see it, is the wide variety of exercise that is possible. A person need never face the bore of monotony from practising one set of exercises. There is plenty to draw from which will provide and keep up a continual interest in your exercise practice. The results obtained become more visible and more rapid. You actually feel you are getting somewhere. Your shoulders will fling back and your breathing will be deeper and more voluminous, the back will flatten—become straight, but do not try to force this straightness. Nature has adapted the muscles to take care of this and will perform the posture naturally without any forced help, as the back muscles build up.

Some men stand so erect they unfortunately fall into the "sway back" position. The hips are carried forward and chest back. Take a fellow like that on a fairly long walk and he will tire out quickly. There is too much strain thrown on the nervous system in the lower back. It is this strangled nerve condition that reduces the muscular strength through fatigue. You will notice all fat men walk with the hips thrown forward. The weijght of their abnormal sized abdomens compels them to walk with the back bent back far beyond the perpendicular line through the center of the pelvis. To arrive at the correct straightness of posture a perpendicular line should be drawn through the middle of the ankle bone, the knee, hips, shoulders and the head. If this line passes through the center of these parts you are O. K. You can easily find out by nailing perfectly plumb a round stick on a flat foot rest and standing sideways to it before a mirror. Your posture will be reflected and by leaning slightly one way or the other you will get the "feel" of the correct carriage for your body, and from this knowledge acquire the habit of controlling your daily movements.

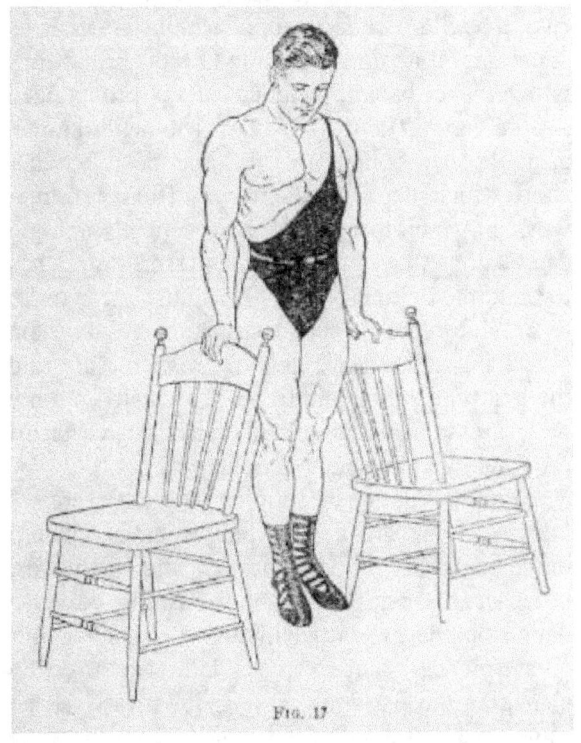

Fig. 17

You rarely see any people more graceful than expert dancers. Posture with them is an art and they spend a great many hours perfecting their erect carriage. Their bodies always carry a well-balanced control from the hips. Still, we cannot all be dancers. I merely refer to them as having the exact, perfect posture. Of course, dancers rarely have a pair of powerful shoulders, but what they have are well moulded. Yet, it is not fair to overlook the fact that not every person finds it so easy to build a broad pair of shoulders; there is sometimes a natural condition in their framework which makes back broadening of any pronounced degree very difficult. The clavicles, better known as the collar bones, are in some people rather short which deprives them of sufficient framework on which to build. Since practically nothing can be done to lengthen them a person simply has to build the deltoids or shoulder muscles, in order to provide more muscular structure. To a certain extent this takes place as you broaden and build the latissimus dorsi and trapezius which are more or less associated with each other by reason of proximity and muscular action.

You will notice by running over the exercises I have so far given that they embrace upward movement and forward, sideways and backwards, but there is one very good pushing exercise which begins as shown in Fig. 17. One of our great pioneers in exercise believed that by clenching the fist and pressing downward, at the same time depressing the shoulder from the erect carriage, sufficient exercise was provided for the latissimus dorsi muscles. There is no doubt it has plenty of merit, but I have found a little variation will give better and quicker results, besides completely eliminating all nervous tension that is created in the first named movement.

Take two chairs and stand them back to back and stand between them with the hand resting on the back of each chair. Stiffen the arms, then draw the heels toward the hips so that you are suspended between the chairs by the support of the arms. From this position you ean commence the exercise. Release the muscular tension of your shoulders and allow y®ur body weight to be lowered so that the head sinks between the shoulders. The shoulders will be hunched up to the ear. Now mind, you do not bend the arms, you merely allow the body to sag between the shoulders; this done —press hard upon the chair backs with the hands and lift the shoulders up as high as you can. This will create a stronger latissimus action. Repeat it as many times as you wish. This simple exercise will give you the same results as are gotten by these experienced athletes who are able to perform the "Full Mount" on the Roman Rings and the "press out" and the "pump exercise" on the parallel bars and the "Body Plant" which is somewhat like a "planche." but the body is held in a line level with the parallel bars, the balance preserved by the hands alene.

Among the present galaxy of body builders, Coulter and Shaffer display about the best type of bade. The beauty of their backs is that they combine beauty, size and strength. You will recall a little earlier in this book I wrote that many had fine appearing backs," but the muscle was devoid of corresponding strength. You would really be amazed if you were to realize the great number of body builders who have a spectacular back development and that is all. It must be an embarrassing situation to be able to show off fine development but not be able to perform feats of strength better than the average person. I would rather not have that type of muscle. I believe you are of the same mind. The trouble with them is their method of exercise has not provided them with sufficient resistance, therefore, they have not pregnated the muscle tissue they have built with the strength ingredients and the ligaments and sinews are just as weak as in the first place.

Many ask me "What is strength? How is it defined?" Most people associate muscle and strength as being inseparable, which is not true—though it can be. It is all in the creation of the muscle tissue.

The finer the muscle tissue is and more tightly it is woven together the stronger the muscle will be. This is only made possible by the fibrous tissue becoming saturated and impregnated by the remarkable chemical ingredients manufactured by the organs. You might liken a muscle to the battery in an automobile. Take two batteries of the same size, same amount of cells and the same amount of liquid, but one is full of power and the other is almost static. The better quality of the one is like the better quality of muscle structure—the quality of chemical within the juice is what gives off more power. It has more electric vibration with which to charge the motor which is similar to the power contained in the high grade muscle which has more nerve vibration because of the chemical properties throughout the muscles and the greater quantity of fibrous tissue provides more positive points for vibration. Their nearness, or the density of this structure, gives shorter trans- mittance, therefore quicker and more dynamic muscular action.

The old manner of practising the exercise last given will give muscle of the spectacular type but the progressive version I have given is what will give you both size, structure, and strength. It not only increases the visible appearance of the latissimus dorsi but by means of the greater effort involved stimulates the superficial muscles of the back. These are the muscles that lie underneath the surface muscles. They are often referred to as "feeder muscles" and I believe there is a great deal of truth in the reference. The superficial muscled are rarely large but what is more important, they are alive with nerve energy which they are capable of pouring constantly into the major muscles according to the demand. Too little attenton is given to these hidden muscles. Many exercise fans will not practice the exercises that increase their muscular vitality. Some who do, practise the exercises like the youngster who practises his music lesson with his eyes on the clock. In this case the fellow who ignores the exercise is the better off of the two.

i have referred to the erector spinus muscles sufficiently to impress you with their vast importance and to carry out my point of the necessity for developing the superficial muscles. I want to go back to them for a moment. You will recall how I explained that these muscles displayed their prominence in the small of the back; leaving this region they cease to become superficial. I do not have to explain to you how foolish it would be to develop the erector spinus muscles in the small of the back and neglect the part of these two rope-like muscles higher up the back. Because you cannot see them does not mean their value to the spine does not exist further than the small of the back. From neck to hips every fraction of an inch is important. Further up the soine they are covered by the latissimus dorsi and trapezius muscles. Near- ing the neck region they_ become more vitally important. Hereabouts centers the most important source of spinal nerve power. The better the condition of the erector spinus the more nerye vibration is stimulated throughout the spine. Bearing this point in mind I would advise you not to be negligent. You are never stronger than your weakest link, and my experience proves to me the weakest link in most body culturists lies in their hidden powers and sources. "Out of sight out of mind" may be a good proverb in some cases, but is never so in the body building sclieme.

I recall a South African story of a man who slaved for years on his tract of land. Finally, becoming discouraged, he sold out and moved away. A few years later he returned ana was amazed to find the tract of land he had owned was now a valuable diamond mine. This poor fellow lacked the perspective or energy to go below the surface. He was much like a great many body builders who go to the theater to see a strong man perform. They come back pepped up with the desire to get a similar body and for a while practise—but only too soon they give it all up as the bunk. They fail to go beneath the surface and mine the nutritive sources ©f energy which lie within us all and draw from these power houses the material to build up worth while muscles.

Perhaps you have noticed nearly all fellows who are round shouldered have a pair of shoulder blades that stick out like sprouting wings. This condition is never seen on a brawny backed man. The "wing" condition is a sure sign of muscular back weakness The shoulder blades are referred to in anatomy as the "weak joints." They lie imbedded within the back muscles loosely and hold their position only by attachment supplied them by the muscles. Naturally this condition is important in order to give the arms and shoulders the free, flexible movement required for our physical action. Development of the upper back muscles tends to shorten them, which is their best natural condition. This is quite the reverse to what nature requires of the leg, arms and stomach muscles. These latter muscles must have long range extension and must normally be extended. But for the back—no; they must be shortened, drawing the shoulder blades together and squaring the shoulders.

Hackenschmidt had a back as big and as flat as a table top and so loose and flexible were these muacles he could shake his back muscles like a dog shakes water off its batk, but when he lurched forward and struck his opponent with his weight and back power the opponent invariably went down as before a battalion of soldiers.

The beauty of back building is that it does so much toward building the chest. In fact, you cannot have a good back without getting a good chest, so chest builders will find it highly profitable to study the cultivation of the back.

I remember giving an enthusiastic back builder the following exercise and his remark after practise was, "Gee, I felt that as """Mich on the chest walls and muscles as upon the back." "Take a dumbbell in one hand and raise it to arm's length overhead. Have the feet spread well apart and hold the other arm outstretched overhead. The object is to toss the dumbbell from one hand to the other a number of times, keeping the arms as straight as possible.

You see the shortening of the back muscles is what widen* the chest, but doa't think all the muscles of the back shorten —it is just those in the upper back and the erector spinus. The latissimus dors! do not so much.

Wheft you nave gotten these muscles all well developed you will be capable of performing many feats of muscle control such as the "spread" and the "roll." Your carriage will be erect and the abdomen will flatten out as the back flattens. The abdominal space of the thoracic arch will widen and give more stomach space, the external oblique muscles of the sides will give you the square waist line back and front. Your entire physical condition will vastly improve. Nerve energy will pour through the spine to feed all the muscles of the body. You will realize how much your arm strength depends upon the amount of back power you have and altogether you will acquire a better understanding of your muscular geography. But don't funk any issues. Do not let yourself be persuaded because you cannot see all that is going on, no results are being gotten. Remember the surface muscles depend upon the powers within—both your organs and superficial muscles are links in the chain which provides superlatively robust health with powerful muscles and endless endurance

Your back contains the spunk to enable you to back the world —if you do not fail it, it will not fail you. It is the back which backs you ALWAYS.

Finally, let me advise you to always, as much as possible, keep a straight back. It is not necessary that you force the condition—that would be unnatural. Many have told me that to walk and sit straight causes a back weariness. I tell them there is no more convincing proof that their back muscles are weak. It is weakness of one group of muscles or another which compels the unnatural positions. The remedy is to get busy right away and build a better back.

When stooping forward to pick up an object from the floor, do not bend over from the waist;

bend the knees and lean forward from the hips. It is a fact that rarely is a woman seen to bend forward from the waist; invariably she will bend the knees with one lower than the other—the lower one being the one corresponding with the arm that is detailed to pick up the object.

People who live in countries where the habit is to carry burdens upon the head have splendid figures. The act compels a bodily centralization. The chim is carried up like that of a person looking far ahead. Incidentally these people have the best records for long life. The vitality accumulated in the neck and back muscles, sup plied by the healthy and powerful condition of their muscles, also prevents the individual from lapsing into unnatural bodily conditions. Remember, avoid unnatural positions, for the moment this happens an unusual pressure is thrown against some internal organ.

Man is a perpendicular creature by nature and it is up to you to carry yourself in the perpendicular as much as possible. Life is extended by keeping yourself spinally young. It is not difficult to do this. You can acquire such a habit easily and it will pay you large dividends in health and strength. After all, the real joy of living lies in having robust health and unlimited energy. Your back is packed with reservoirs of energy It is up to you to develop this energy and feed the muscles of your body from these vital sources.

www.ingramcontent.com/pod-product-compliance
Lightning Source LLC
Chambersburg PA
CBHW060229290526
45789CB00003B/1476